A Look At
ROTATIONAL MOTION

Rebecca Woodbury, Ph.D., M.Ed.

Gravitas Publications Inc.

A Look At
ROTATIONAL MOTION

Illustrations: Janet Moneymaker

A Look at Rotational Motion
ISBN 978-1-950415-24-3

Published by Gravitas Publications Inc.
Imprint: Real Science-4-Kids
www.gravitaspublications.com
www.realscience4kids.com

RS4K

Photo credits: Cover & Title Page: By JenkoAtaman, AdobeStock; P.3. By Tomsickova, AdobeStock; P.5. By Joanna Zielinska, AdobeStock; P.9. By JenkoAtaman, AdobeStock; P.12. By Pexels from Pixabay

What happens when

you ride a bike?

You go
fast?

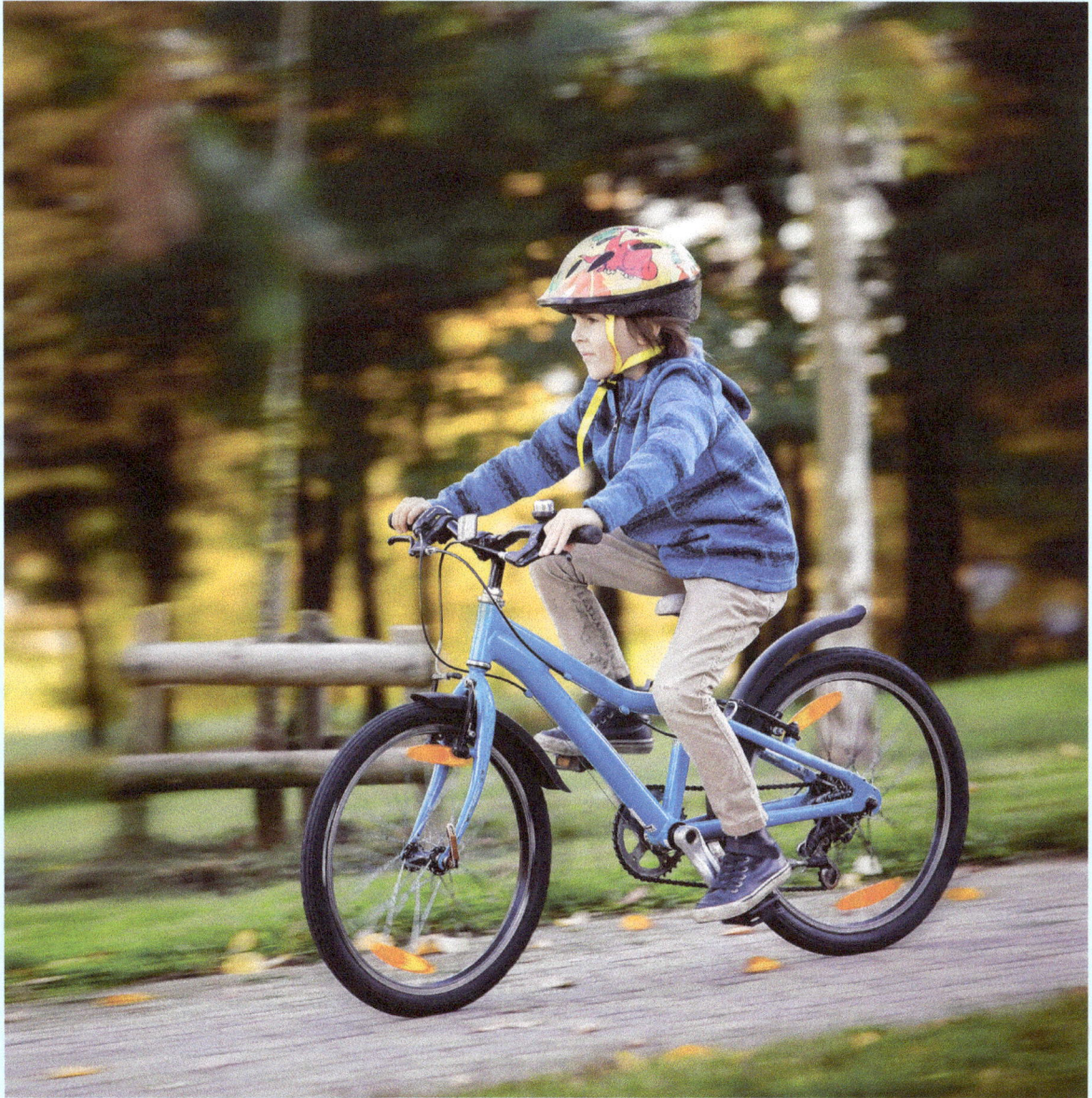

When you push on the pedals,
the bike moves forward.

A → B

You may know that when your bike moves in a straight line, you and the bike are using **linear motion.**

Review: LINEAR MOTION

The word **linear** means "in a line."

Any object moving "in a line" has

linear motion.

The wheels on your bike
also use linear motion.

I want
a bike!

As you move the pedals,
the wheels move
forward in a straight line.

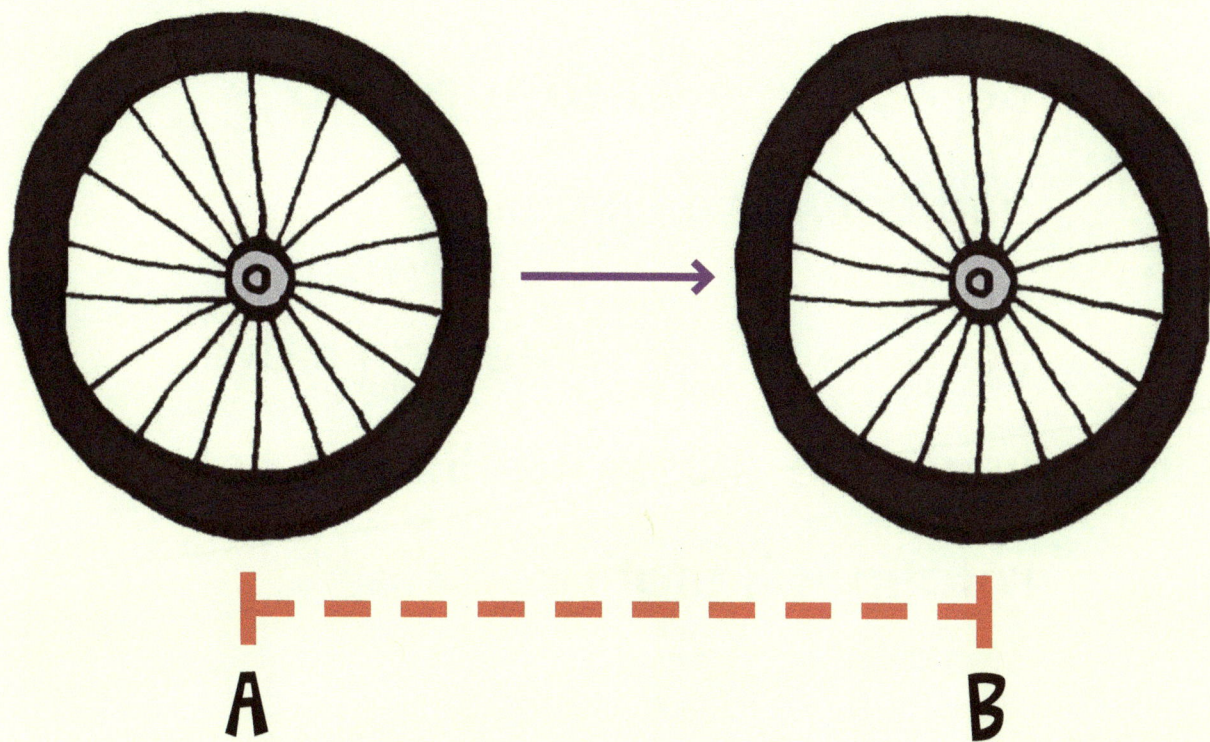

A   B

But a bicycle wheel is **round** and can move in a circle.

Because a bike wheel can move in a circle, it uses **nonlinear,** or **rotational, motion.**

But what is rotational motion?

(Turn the page.)

The word **rotate** means to turn around a center point.

Rotational motion is a type of movement where something turns around a center point.

Center point

To see how a bicycle wheel moves with rotational motion put a dot on one side of the wheel.

When you turn the wheel,
you will see the dot rotate.

Now I know about rotational motion!

Me too!

Lots of things move

using rotational motion.

A spinning top

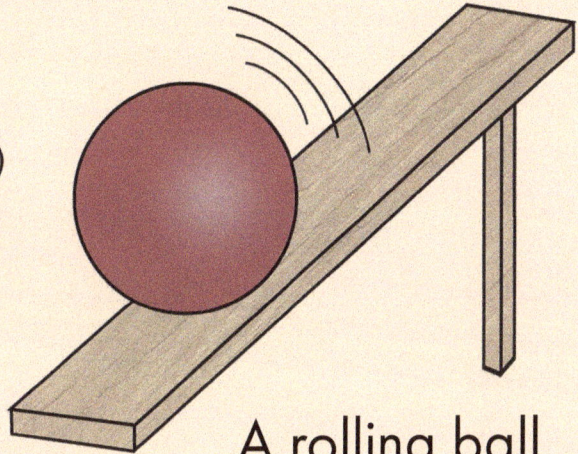

A rolling ball

A windmill

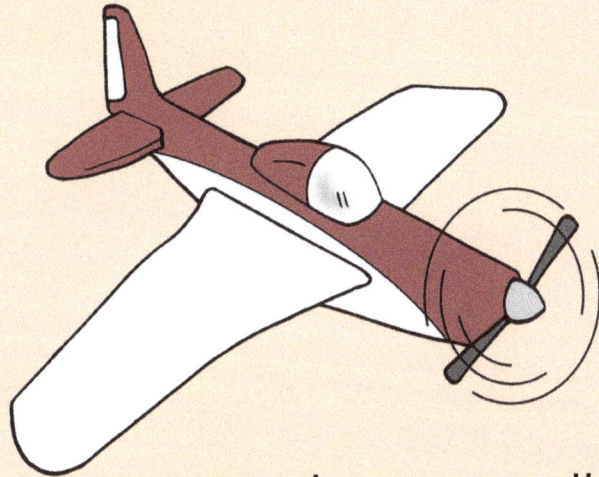

An airplane propeller

How to say science words

circle   (SUHR-kuhl)

linear   (LIH-nee-uhr)

motion   (MOH-shuhn)

nonlinear   (nahn-LIH-nee-uhr)

rotate   (ROH-tayt)

rotation   (roh-TAY-shuhn)

rotational   (roh-TAY-shuh-nuhl)

round   (ROWND)

science   (SIY-ens)